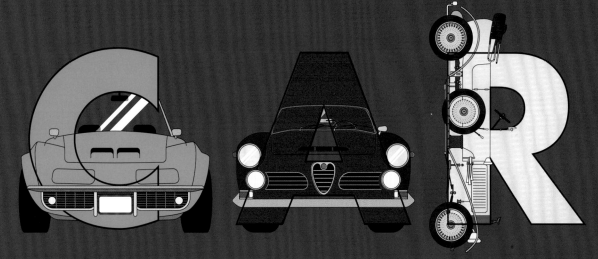

LEGENDS
ALPHABET

Words by Robin Feiner

A is for **A**lfa Romeo. With curvaceous, avant-garde Pininfarina designs, this carmaker brought gorgeous Italian style to the world's roads. It didn't take long before Alfa lit up the track, famously winning the first Formula 1 World Championship in 1950. Even Enzo Ferrari got his legendary start steering Alfa to victory.

B is for **B**MW.
Cars began rolling off the Bayerische Motoren Werke production line in 1928. Since then, 'Beamers' have earned a reputation for engineering excellence and quality. Success in F1 and endurance racing has made BMW M-Series cars some of the most revered drivers' machines of choice—just ask James Bond.

C is for Chevrolet.
Born in Motor City, USA,
Chevy's legendary bow tie
logo has stood for power
and reliability for over
a century. From its sleek,
suave Camaros and
El Caminos to its Corvette
sportscars and rock-solid
454 SS and Silverado trucks,
Chevrolet is a force on the
road and track.

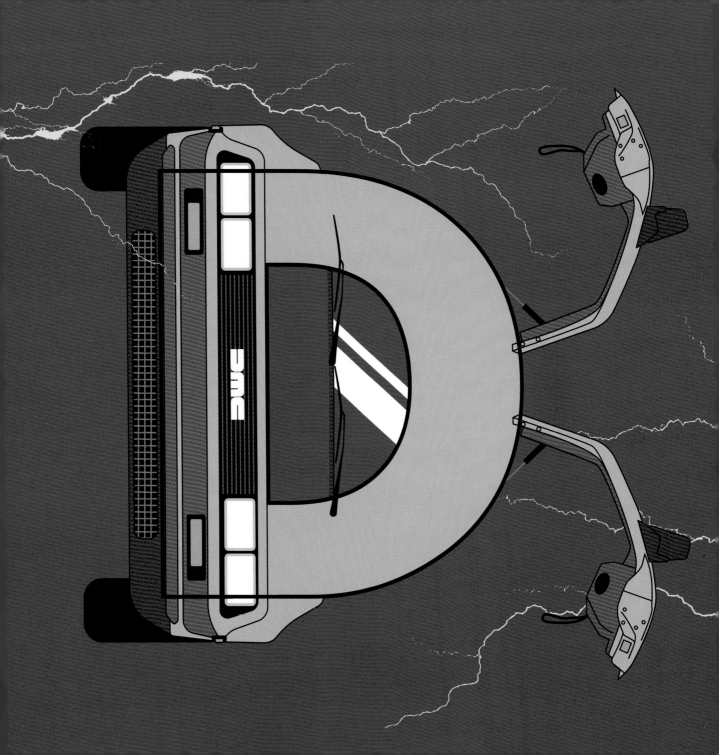

D is for **D**eLorean.
Great Scott! This ride took Marty McFly and fans Back to the Future in 1985. But the DeLorean DMC-12 actually debuted in 1981, impressing many with its legendary winged doors and striking stainless steel body. Too bad production ended the next year.

E is for Eunos.
The Eunos Roadster,
also known as the Mazda
MX-5 Miata, was born in
the 1990s. It's a front-engine,
rear-wheel drive, lightweight
roadster like the famed
Lotus Elan. It's now in its
fourth generation and is
the best-selling two-seater
convertible sports car ever.

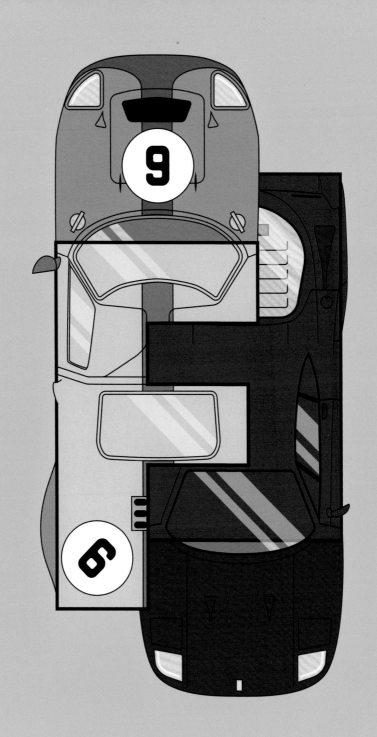

F is for **F**ord vs. **F**errari. Their Le Mans rivalry is arguably the most legendary in automotive history, but that's just part of the story. Henry Ford transformed automobile manufacturing with the famous Model T. Then Enzo Ferrari, with his hand-built masterpieces presented in iconic 'Rosso Corsa,' redefined speed and desire worldwide.

**G is for GMC.
The biggest and meanest
machines to rumble out
of the legendary General
Motors, GMC trucks have
provided groundbreaking
innovations since 1902.
With a range that spans
military support vehicles
to cross-country shipping,
GMC machines, such as the
legendary Hummer, are
a seriously big deal.**

H is for **H**onda.
Its high-revving engines
famously powered Ayrton
Senna and the mighty
McLaren to victory in F1.
It also put this talent on the
road with the legendary
NSX, beating Ferrari at its
own game. Honda has always
proudly helped Japan put its
best foot forward in the auto
world, dominating MotoGP
as well.

I is for **I**mpreza.
Behind the punching pistons of its legendary turbo boxer engines and all-wheel drive, Impreza triumphed in the WRC in the mid-'90s and early 2000s and has emerged victorious atop Pikes Peak on many occasions. With its rock solid off-road reputation, it's no surprise Subaru has become the go-to Japanese adventure brand.

J is for **J**aguar.
Revered for its '50s race cars
wearing British Racing Green,
and its luxurious saloons,
they're the vehicle of choice
for the British Royal Family.
One of Jaguar's greatest
moments came with the 1961
E-Type that wowed audiences
at the Geneva Motor Show.
It's now considered one
of history's most beautiful
and legendary designs.

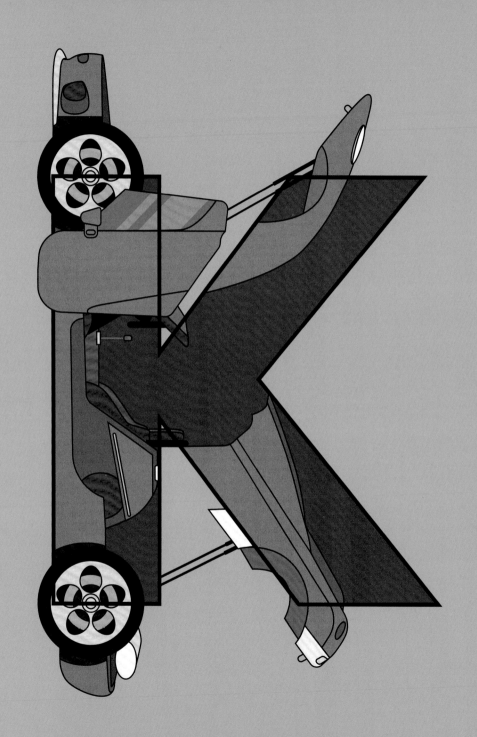

K is for **K**oenigsegg. Unique, exclusive, innovative, and mind-bogglingly fast, these Swedish craftsmen have been offering hyper-cars to the hyper-rich since 2002. These legends created the world's fastest production car in 2017, capable of record-setting acceleration from 0-400 km/h before you finish reading the next paragraph.

L is for Lamborghini.
In the '60s, Lamborghini
invented the supercar with
the stunningly gorgeous
Miura V12. And the legendary
'70s Countach, with its scissor
doors and Italian Wedge, is
possibly the most astonishing
four-wheel design of all time.
Driven by passion and power,
Lambos are the flashy poster
boys of speed.

Mm

M is for Mercedes-Benz. Many call Germany's Karl Benz the inventor of the gas-powered automobile. And the marque bearing this legend's name has continued to innovate for more than 135 years, from the 300SL Gullwing and the celebrated Silver Arrows to its saloons and championship-winning F1 machines... and everything in between!

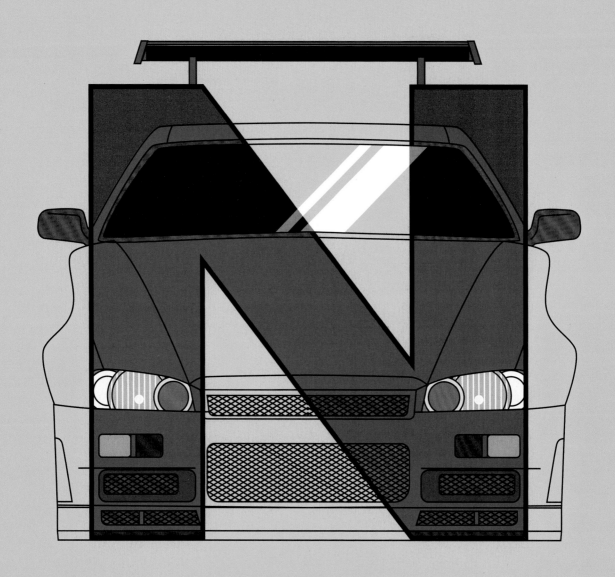

N is for **N**issan.
This Japanese brand's
Skyline GT-R took on the
Porsche 959—and won.
With supreme handling
and performance, it turned
in the first sub-8-minute
production car lap at the
iconic Nürburgring in 2002.
Its big screen and video
game fame make this fast
and furious machine a legend
amongst drifting diehards.

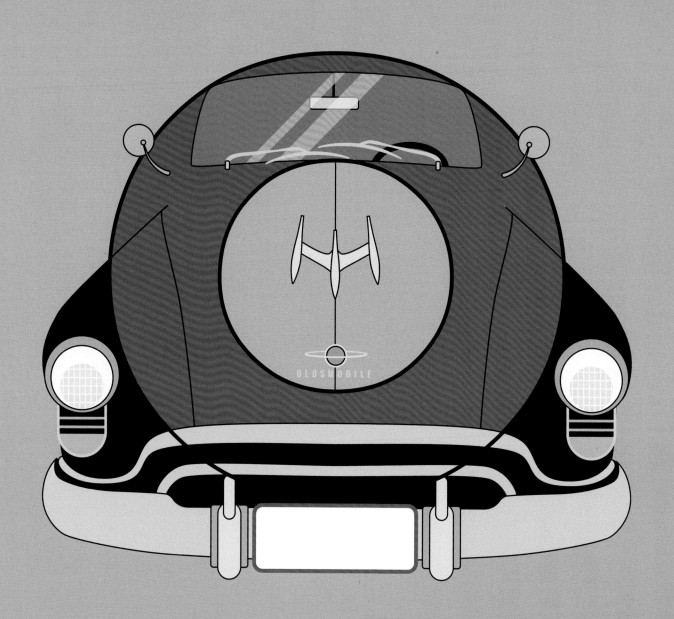

Oo

O is for Oldsmobile. This legendary marque from Michigan helped power America's 20th-century car boom. Innovative and influential in its day, we can thank Oldsmobile for chrome trim, automatic transmission, V8 engines, and the first mass-produced turbocharger. No wonder they had such success in NASCAR and at the Indianapolis 500.

P is for **P**orsche. From solid beginnings with the tiny 356, Porsche has become one of history's most important automotive brands. Crushing dominance in motorsport's most grueling event—the 24 Hours of Le Mans—has come from unmatched dedication to engineering advancement. However, the iconic lines of the legendary 911 have hardly changed since 1963.

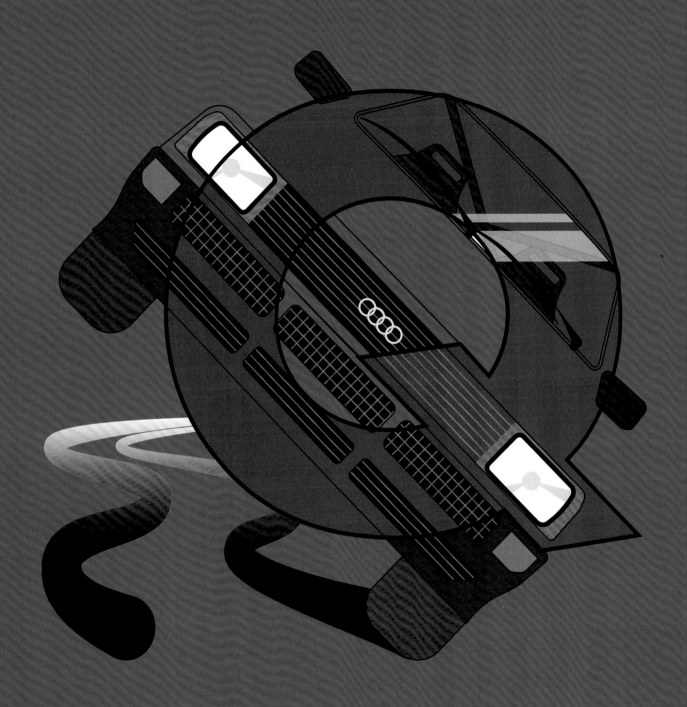

Q is for Audi Quattro. The legendary Audi all-wheel drive system, known as Quattro, offered superior traction and helped Audi revolutionize rally racing in the 80s. Iconic wins at Pikes Peak followed before Audi introduced AWD to road cars to improve handling and safety. Major manufacturers, such as Subaru, have followed in their tracks.

R is for **R**olls-Royce. Charles Rolls and Henry Royce started creating luxurious vehicles of elegance and grace in 1906. With the finest leather and wood, and engines that purr, it was said that at 60mph in a Rolls, the loudest noise comes from the electric clock. The silver-winged lady adorning every hood is motoring's most legendary mascot.

S is for **S**helby.
Founded by Le Mans champion Carroll Shelby, this American marque is all about speed. Shelby brought the British roadster style to the US, adding small-block American engine power to create legendary models such as the Cobra, the GT500, and the Eleanor—the car made famous by Gone in 60 Seconds.

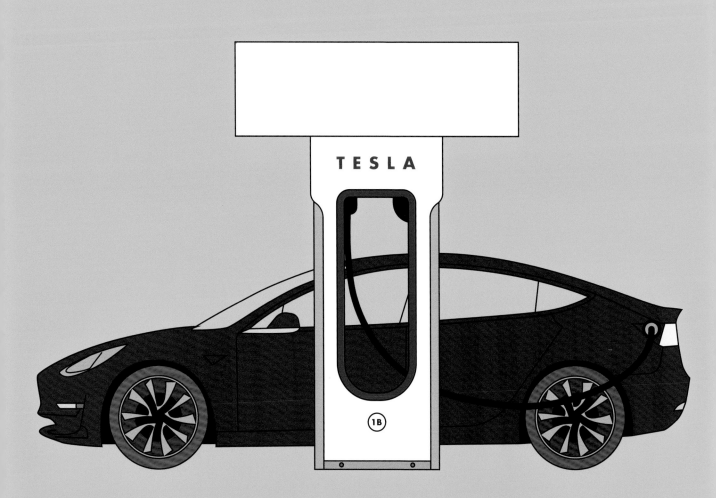

T is for Tesla.
Eco-warrior Elon Musk is leading the charge in making electric vehicles stylish, reliable, affordable, and practical. Going green has never been more attractive! The Model Y became the world's bestselling car in 2023. At the same time, Tesla was the fastest-growing automotive maker.

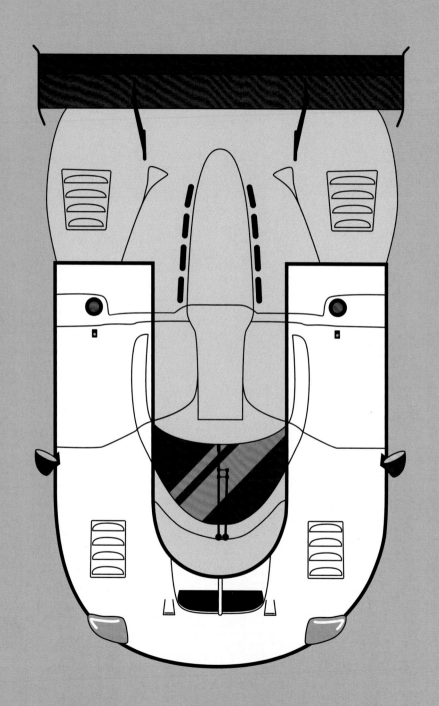

U is for **U**ltima.
Starting with the Mk1 in 1983, Britain's Ultima cars were inspired by lightweight, aerodynamic Le Mans machines. Customizability makes them unique, while blisteringly powerful Chevy V8 engines have helped achieve legendary record-breaking speeds. Order yours as a kit, roll up your sleeves, and build it yourself!

**V is for Volkswagen.
'The People's Car,'
designed by Ferdinand
Porsche, was originally
produced as an affordable
German family vehicle.
And the legendary Beetle,
with its rear engine, rounded
body, bright colors, and Punch
Buggy nickname, sold more
than 23 million units. Now,
Volkswagen is one of the
biggest auto manufacturers
in the world!**

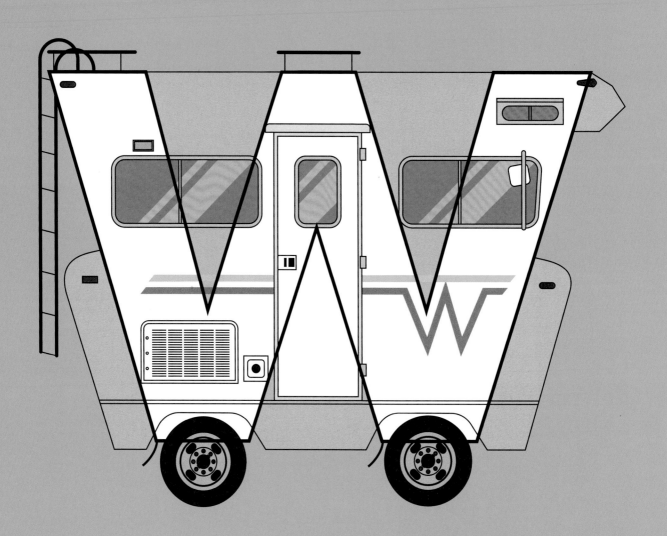

W is for **W**innebago. The biggest name in Recreational Vehicles (RVs) since its creation in 1967, Winnebago has been on the cutting edge of RV innovation, bringing the creature comforts of home to the highway. Its legendary Flying W emblem symbolizes the freedom of the great American road trip.

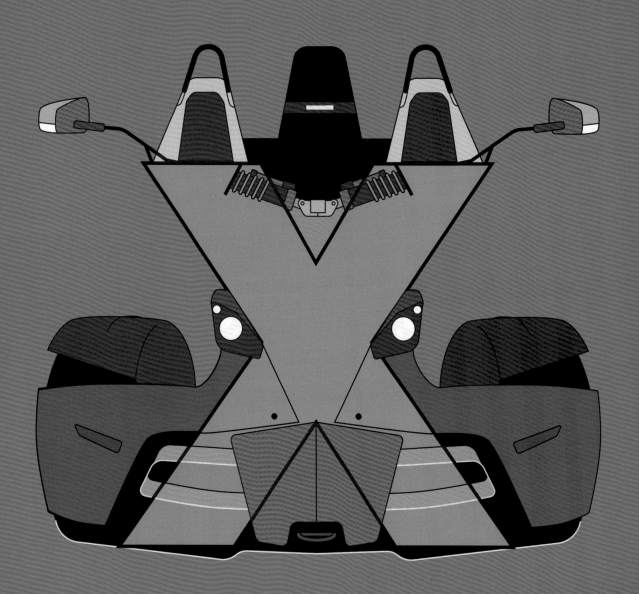

X is for KTM X-Bow. This exciting Austrian motorcycle manufacturer took all their smarts and passion and applied it to a car. The result? A light, agile, radically-designed open-wheeler for the open road. X-Bow, pronounced crossbow, with its raw, race car performance, reminds enthusiasts of the legendary Lotus 7.

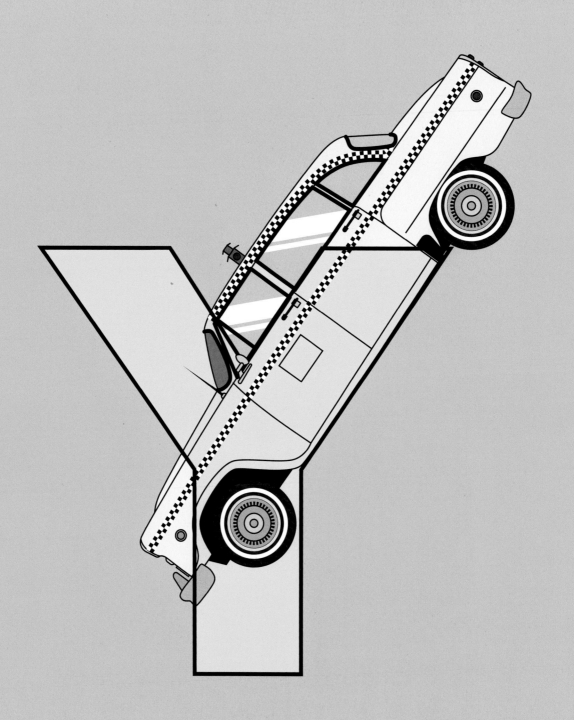

Y is for **Y**ellow Cabs.
The Chicago-based Yellow
Cab Company cruised to
popularity in the early 1900s.
Owner John D. Hertz made
his fleet yellow so they were
easier to spot in the distance,
but his sights were set on
the stars. Yellow became the
official color of legendary
New York taxis, and Yellow
Cabs became one of the
biggest taxicab operators
in the US.

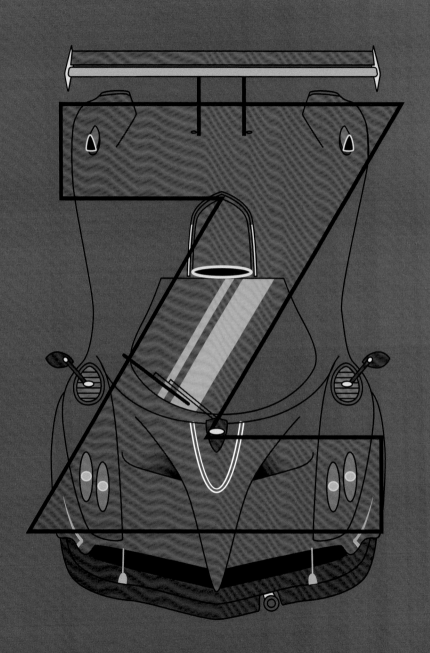

Z is for Pagani Zonda. With experience at Lamborghini, Horacio Pagani set out to create an Italian marque focused on design, innovation, power, and craftsmanship. After seven years of development, the breathtaking Zonda was born. With its carbon fiber body and Mercedes-AMG engine, it's an automotive masterpiece and a true legend.

The ever-expanding legendary library

EXPLORE THESE LEGENDARY ALPHABETS & MORE AT WWW.ALPHABETLEGENDS.COM

CAR LEGENDS ALPHABET
www.alphabetlegends.com

Published by Alphabet Legends Pty Ltd in 2023
Created by Beck Feiner
Copyright © Alphabet Legends Pty Ltd 2023

Printed and bound in China.

9780645851427